River in the Ocean
The Story of the Gulf Stream

by Alice Gilbreath

Dillon Press, Inc.   Minneapolis, Minnesota 55415

To my son, Rex

Library of Congress Cataloging in Publication Data

Gilbreath, Alice Thompson.
  River in the ocean.

  Bibliography: p.
  Includes index.
  Summary: Describes the warm river flowing in the Atlan-
tic Ocean, the studies that have been made of it, and its
current and future importance.
    1. Gulf Stream—Juvenile literature. [1. Gulf Stream]
I. Title.
GC296.G55   1986   551.47'1   85-6883
ISBN 0-87518-297-6

Dillon Press, Inc., 242 Portland Avenue South
Minneapolis, Minnesota 55415

Printed in the United States of America
2 3 4 5 6 7 8 9 10 94 93 92 91 90 89 88 87

Contents

Ocean World Library

 Gulf Stream Facts

Size:
>   Varies in width from about 40 miles (65 kilometers) at the Straits of Florida to several hundred miles as it becomes part of the North Atlantic Current. Transports more water than all of the rivers in the world combined

Temperature:
>   Ranges from 11 to 18°F (6 to 10°C) warmer than water around it; varies from 54°F (12°C) to 75°F (24°C) from one side of the current to another in the North Atlantic

Depth of Heat:
>   At equator, heat penetrates 100 yards (91.44 meters) below the surface into the North Equatorial Current; the Gulf Stream is formed out of this current

Distance Its Heat Is Transported:
>   From the Gulf of Mexico to Iceland, Norway, the British Isles, and other European coastal regions

Speed:
>   At Straits of Florida—about 5 miles (8 kilometers) per hour
>   After cooling down and spreading out—about 3 miles (5 kilometers) per hour

Swiftest Part of Gulf Stream:
Usually 12 to 25 miles (20 to 42 kilometers) from the west side of the stream

Weather Making:
Partly responsible for many hurricanes and some tornadoes; moderates the weather of Iceland, the British Isles, and other areas of northern and western Europe

Named By:
Benjamin Franklin

First Measured By:
Lieutenant J. E. Pillsbury

First Research Submarine to Drift in the Gulf Stream:
Ben Franklin (1969)

First Viewed from Space By:
A U.S. satellite, *NOAA-5* (1976)

Specializes in its Study:
Woods Hole Oceanographic Institution in Massachusetts

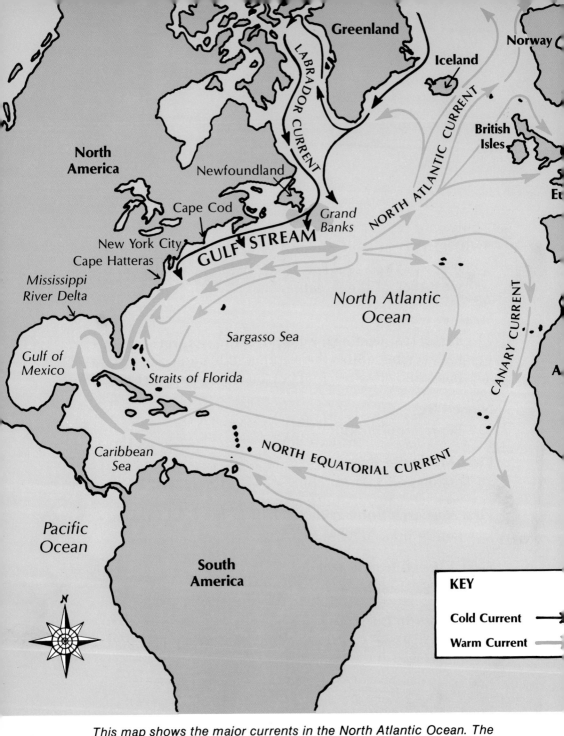

This map shows the major currents in the North Atlantic Ocean. The Gulf Stream and other currents form a circular pattern of warm, flowing waters around a large area of the North Atlantic.

 # The Big Ocean River

The **Gulf Stream**,* a famous and unusual river, flows in the Atlantic Ocean. Unlike other rivers, it never dries out and never causes floods by overflowing its banks.

It is difficult to imagine a river in the ocean because a river must have banks and a riverbed. Clay, sand, and rock form the banks and riverbeds of most land rivers. The Gulf Stream's banks are layers of cold ocean water on both sides of this big river. These cold water banks are more flexible than the earthen banks of land rivers, but they serve the same purpose. They form a channel in which the river can flow. Underneath the Gulf Stream a cold current flows south from the **Arctic Circle**. This current flows in the opposite direction from the Gulf Stream and forms a riverbed for the big ocean river. Because warm water of the Gulf Stream **expands** and rises, it flows above the heavier cold water that **contracts** and sinks.

The Gulf Stream and the cold current also contain

*Words in **bold type** are explained in the glossary at the end of this book.

different amounts of salt. Because the salt content of seawater makes it more or less dense, the ocean river and cold current do not have the same **density**. And, since water of different densities does not mix rapidly, the Gulf Stream and the current tend to stay apart.

A Thousand Mississippi Rivers

The Gulf Stream transports more water than all of the rivers in the world combined. Through the Straits of Florida, at the beginning of its journey, it pushes millions of tons of water per minute.

Imagine the Gulf Stream as a land river in the United States.

If you have seen the Mississippi River, you know that it is a huge river—the largest in North America and one of the biggest in the world. Ocean vessels sail in the Mississippi River because it is wide and deep. Yet it would require more than one thousand Mississippi rivers to carry the water transported by the Gulf Stream.

If it were possible to squirt all of the water from the Gulf Stream through a giant fire hose onto North America, the entire continent would become a wading pool in one day. This giant wading pool would be

A fishing boat speeds out into the Gulf Stream near the coast of Florida.

warm, for the surface of the Gulf Stream ranges from 11 to 18°F (6 to 10°C) warmer than the water around it. At its beginning, or source, in the western Caribbean Sea near the Gulf of Mexico, it is about the temperature of a heated swimming pool.

Energy from the Sun

Just as the sun provides **energy** for all life, it provides energy that warms the Gulf Stream. The North Equatorial Current, out of which the Gulf Stream is formed, absorbs the sun's heat at the **equator**. Here, a great deal of heat is concentrated in a small area. In fact, so much heat is absorbed that the water is warmed to nearly 100 yards (91.44 meters) below the surface.

Warm water at the equator expands, making the sea level a few inches higher. This rise produces a tiny slope so that seawater flows from the equator "downhill" toward the poles. Water that becomes the Gulf Stream flows toward the North Pole. It becomes part of a large, clockwise system of currents in the North Atlantic Ocean.

This big river in the ocean moves at different speeds in different parts of the Atlantic Ocean. At first, when it is loaded with heat from the tropics and forced through the narrow Straits of Florida, the Gulf Stream moves at about five miles (eight kilometers) per hour—about the speed of a freestyle swimmer's sprint. As it flows onward, spreading out and **radiating heat** as it goes, some of its speed is lost. Finally, when it has cooled down and spread out considerably, it flows at less than three miles (five kilometers) per hour.

The earth's **rotation** helps move this big river. Its spin, called the **Coriolis effect**, causes the curve of the Gulf Stream and other moving objects. The spinning earth moves them clockwise in the Northern **Hemisphere** and counterclockwise in the Southern Hemisphere. It helps push the Gulf Stream around the Atlantic Ocean in a circular, clockwise pattern.

Blowing in the Wind

Winds, set in motion by the earth's rotation, also play a part in moving this big river. To understand how they do, blow on a pan of water. Even gentle blowing sets the water in motion.

Trade winds (meaning steady winds) blow westward toward the equator in both hemispheres. Then

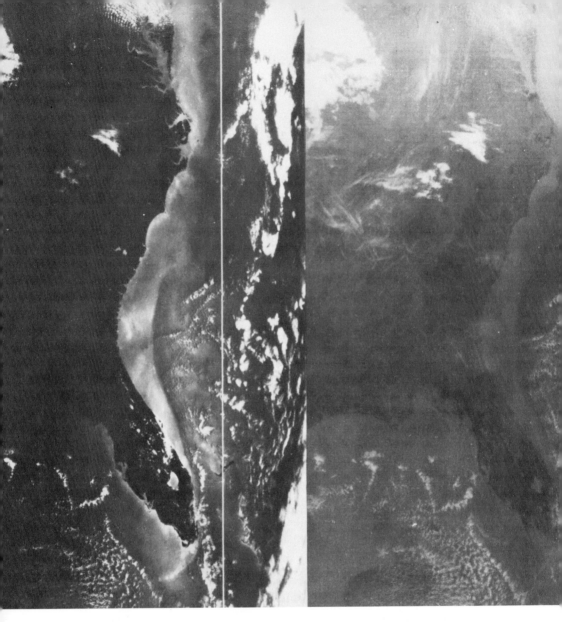

These two photographs of the southeastern United States were taken from space by the NOAA-2 satellite. The photo on the right shows visible, reflected light. The one on the left is an infrared photo that shows temperature differences in the earth's land, ocean, and clouds. In this photo the Gulf Stream is the dark, winding band that appears a short distance from the U.S. east coast.

they curve toward the northeast in the Northern Hemisphere and toward the southeast in the Southern Hemisphere, pushing the surface currents in their paths. They push water that will become the Gulf Stream into the Gulf of Mexico. Then westerlies (winds that blow from west to east) help move the Gulf Stream up the east coast of the United States and across the Atlantic Ocean.

Because of its flexible banks, the Gulf Stream does not always flow in exactly the same place in the ocean. Its position changes by seasons and sometimes changes several times in one day as the big river **pulsates** from side to side. Over a period of a few months, its banks may shift one direction or another as much as 250 miles (400 kilometers).

It's not surprising that sailors were terrified when they first sailed into this powerful current. Many were superstitious. Some believed that evil spirits could attack ships and cause all sorts of disasters. Sometimes a ship's captain realized his ship was sailing too fast for the wind that pushed it. Because of his own superstitions or for fear of mutiny from a superstitious crew, he made false entries in his log book, showing a slower speed.

From the Gulf to Iceland

For centuries people have thrown bottles into the oceans with notes inside showing the date and place they entered the water. There is something fascinating about watching a drifting bottle and wondering where it will go before it finally washes up on a beach. Some of these bottles drifted for years, pushed along by first one current and then another, before they washed ashore.

A Bottle in the Ocean River

Imagine that we throw a bottle into the center of this big ocean river at its source, in the western Caribbean Sea. If we follow the bottle, it will show us where the Gulf Stream flows.

The bottle will be pushed rapidly through the forty-mile-wide Straits of Florida. Like land rivers, the Gulf Stream moves faster when its channel is narrow. Also like land rivers, the speed of this big river is faster near its center.

Near land, the bottle will float in the Gulf Stream northward up the east coast of the United States. It will float past Florida, Georgia, South Carolina, and on toward Cape Hatteras, North Carolina. With well defined banks and a riverbed of colder ocean water, this portion of the Gulf Stream is a true river in the ocean.

Beyond Cape Hatteras, the Gulf Stream begins to widen. As it spreads out, the big current and the bottle slow down slightly. The bottle will continue floating up the coast, riding the crest of the current driven by heat, wind, and the earth's spin.

Probably the bottle will stay in the main flow of the Gulf Stream. However, it could be swept off in one of the **whirlpools**, or **eddies**, that break off from the main current. Eddies whirl in a counterclockwise direction away from the Gulf Stream and often surround a pocket of cold water. If the bottle is caught in an eddy, it will be whirled away for many miles. When the eddy loses its energy, the whirling stops, and its water combines with ocean water around it. Then the bottle will drift in the ocean until waves bring it to shore or until it moves again into the path of the Gulf Stream.

In this photograph, taken from the NOAA-7 satellite, the Gulf Stream appears as the dark band at the left of the picture near the U.S. east coast. As it moves away from Cape Hatteras and the coast, the Gulf Stream begins to widen in the North Atlantic Ocean.

Let's assume that the bottle is not swept off, but stays in the Gulf Stream. It will float a little farther from the coastline and continue more or less northward as far as New England. Then the earth's west-to-east rotation moves the Gulf Stream across the Atlantic Ocean in a northeasterly direction.

Where Warm Water Meets Cold

Near Newfoundland the bottle will be tossed and spun and joggled by **turbulent** waters. Here the Gulf Stream collides with the **Labrador Current**, a powerful, icy flow from the Arctic Circle. The bottle will be washed this way and that for hours or days before it floats off again with the warm current.

Because of this collision with the cold Arctic current, the Gulf Stream becomes cooler. Now the stream spreads out until it is much wider and slower and its banks are less stable. The mighty ocean river is losing some of the energy that made it so powerful. While it is usually known as the Gulf Stream all the way across the Atlantic Ocean, it is now a part of the North Atlantic Current.

The Gulf Stream also curves from side to side, snake fashion. Now that its banks are beginning to collapse, cold water somewhat penetrates this mighty current from the right side.

If the bottle remains in the Gulf Stream, it will continue across the Atlantic Ocean. Before it reaches Europe, the widening current's banks become even less definite. It has changed from the narrow, ribbon-like stream that poured through the Straits of Florida to a huge ocean river several hundred miles in width. Even so, it has kept some of its warmth.

Along the way, cold water from the north divides the stream so that it develops more than one branch. Perhaps the bottle will remain in the branch—still called the North Atlantic Current—that flows toward Iceland, Norway, and the British Isles. If an eddy sweeps it out of the stream here, the bottle may be cast upon the shore of one of these countries. If it drifts with another branch of the Gulf Stream, it will travel toward Spain and then south along the western coast of Africa. Along the way, it joins the Canary Current.

Adrift in the Sargasso Sea

The Gulf Stream, along with yet another branch of the North Atlantic Current, circles clockwise around the almost motionless **Sargasso Sea**. The cur-

A "hot water tank" in the Atlantic Ocean, the Sargasso Sea is covered by millions of tons of floating seaweed.

rent becomes a boundary for parts of this big sea which serves as a "hot water tank" in the Atlantic Ocean. The Sargasso Sea is covered by millions of tons of floating seaweed. Berries on the seaweed serve as **pontoons** to keep the plants afloat. The seaweed is so thick that, from a distance, the Sargasso Sea appears to be an island.

Long ago sailors were superstitious about the Sargasso Sea. They called it the "Sea of Mystery" or the "Coagulated Sea." Because of its dense plant life, they believed that a ship could get so entangled in this ocean jungle that it would never get out. Since the Gulf Stream's swift current bordered this quiet sea, sailors thought that a ship could be swept out of the current and into this watery jungle. Possibly this superstition lived on because ships were at the mercy of winds. A ship in the Sargasso Sea would not have the help of a strong current or a strong wind to move it on.

If our bottle drifts into the Sargasso Sea, it may be caught in the warm, watery forest of seaweed and stay there for weeks. Some of this seaweed put its roots into the Sargasso Sea centuries ago. As the seaweed multiplied, ocean animals found shelter there.

The Sargasso Sea serves as a hatchery and nursery for many sea creatures such as the flying fish shown here.

The Sargasso Sea serves as a hatchery and nursery for Gulf Stream creatures. Flying fish build their nests and lay eggs in the seaweed. Sea horses and crabs cling to the floating mass. Small fish swim among the seaweed for protection.

Eventually our bottle will drift from the Sargasso Sea into the path of an ocean current and will be carried away. Ocean currents carry water back and forth between the equator and the poles through all the earth's oceans. Warmer water rises. Colder water sinks, and its place is taken by warmer water. As the Gulf Stream water loses its warmth, it sinks and flows back to the equator. There it heats, rises, and begins its journey northward again. Finally the bottle, too, may complete this clockwise circle, or it may be washed up on a beach along the way.

# Benjamin Franklin Solves a Problem

In the fifteenth century, Christopher Columbus found bamboo and coconuts that the Gulf Stream had transported to European shores. He realized there must be land to the west and a strong current flowing from west to east which was carrying them. Columbus set out to find the New World on his historic 1492 voyage.

In the sixteenth century, a Spanish explorer, Ponce de León, sailed west from Puerto Rico searching for the Fountain of Youth. Along the way he encountered a current so strong that he could not sail against it even with the aid of a strong wind. Slowly his ship was driven backward. He was sailing against the Gulf Stream current.

By the middle of the eighteenth century, sailors were learning more and more about the oceans. Although they learned to sail largely by trial and error, some understood that the ocean had currents and that some currents were larger and more powerful than others.

A Mystery to Solve

Many ship captains had learned to cope with the currents, but the captains of the British mail ships, or packets, sailing from England to New York had not learned. The packets took two weeks longer for the journey than did merchant ships sailing from England to Rhode Island. The British Board of Customs wanted to know why and what could be done about it. It became the job of Benjamin Franklin, who was then postmaster general of the American colonies, to furnish the board with answers.

Franklin considered the problem. Since the merchant ships were heavily laden, they moved slower. Also, they traveled a greater distance. It would make sense for them to take more time—not less—to cross the Atlantic Ocean. What could cause the mail packets to be slower?

Ships sailing from one port to another along the Atlantic coast of the United States encountered the same problem. It took ships three or four times as long to sail from north to south as it did on the south to north return trip. Both the British mail packets and the Atlantic coast ships were being slowed down by an unknown force in the ocean.

Benjamin Franklin solved the mystery of the Gulf Stream and learned that this current was different from any other in the North Atlantic Ocean.

The Whalers from Nantucket

Benjamin Franklin discussed the problem with a Nantucket whaling captain. From the whaler he learned about a wide, swift current in the ocean. The whaling captain told him the current flowed from south to north along the coast, and then crossed the ocean toward Europe in a northeasterly direction. Whaling captains knew a great deal about this swift stream and often crossed it in pursuit of whales. Baleen whales liked to eat plankton from the swift current but did not like the stream's warm water. They stayed in the cold water on both sides of the Gulf Stream where they could eat plankton from its edges.

Whalers told Franklin that ships sailing against the current would be slowed down. Since the current crossed the Atlantic Ocean in a northeasterly direction, ships sailing west in it would have to buck the current all the way across the ocean. The thing for southwest-bound ships to do, the whalers advised, was to cross this powerful stream as rapidly as possible and sail outside of it. This route would be much easier and would save a great deal of time, since the ships would not be sailing against the current. When ships sailed toward England, they should take advan-

Nantucket whalers knew about the Gulf Stream because they often crossed it in pursuit of whales.

tage of the powerful current and sail in it.

These same whaling captains had already informed captains of the British mail packets about this river in the ocean. The British captains, however, had shrugged it off. They thought, how could a whaler know more about the ocean than a British sailing captain?

Whalers advised Benjamin Franklin that merchant ships sailing from England to Rhode Island knew about the swift current. To make the fastest journey possible, the merchant ships were avoiding the current on their trip from Europe to America. On their return trip, they were sailing in the current and receiving a boost from it.

After much checking, Franklin concluded that this powerful current made the difference in the length of time ships took to cross the Atlantic Ocean. Ships sailing in the direction the current flowed were helped to go several miles per hour faster. Ships sailing against this strong current were slowed down by several miles per hour. In a single journey, this difference could add up to as much as two weeks sailing time.

The problem could be solved by finding out exactly where the strong current flowed all the way

across the Atlantic Ocean. Then this information could be passed along not only to the British mail packets, but also to all ships sailing in the Atlantic Ocean.

Locating both edges of this swift current all the way across the ocean was a big job. Again Benjamin Franklin turned to whalers and pieced together the information they furnished. Whalers might be called the first school of American navigators, for they gladly shared their knowledge with American shipmasters and with all others who were interested. Of course, neither Franklin nor the whalers had any way of knowing that the banks of this big river often shifted.

Franklin decided that a **thermometer** would be useful in locating the exact edges of the Gulf Stream. A bucket could be lowered into the ocean from a ship, hauled up, and the temperature checked. Since this current was warmer than the ocean on either side of it, taking water temperatures would help sailors know whether they were in or out of the stream. Air temperatures would be useful, too, since air above the warm current would be warmer than air above ocean water. Franklin took water and air temperatures every time he sailed. He urged others to do the same and share their information.

A Name for the Ocean River

Since Franklin believed this big ocean river began in the Gulf of Mexico, he named it "Gulf Stream." Using information he had received from whaling captains—the most reliable information available—and what he had learned from thermometer readings, Franklin had a chart made of the Gulf Stream showing its size, location, and swiftness. It was a remarkably accurate chart for 1769.

He sent this chart to the British postmaster general, along with directions for avoiding the Gulf Stream on trips from England to New York. He also explained how to take advantage of the swift current on return trips.

However, his charts and his ideas were rejected by the British. Franklin said nothing more to the British about the Gulf Stream until after the American colonies' War of Independence.

Throughout his lifetime, Benjamin Franklin continued to be interested in the Gulf Stream. Each time he journeyed in the Atlantic Ocean, he drew up

Benjamin Franklin's chart of the Gulf Stream was remarkably accurate for 1769. It showed the Gulf Stream's size, location, swiftness, and pattern of circulation in the Atlantic Ocean.

wooden kegs of water and plunged a thermometer into them. From these readings he learned more and more about the location of the Gulf Stream.

Finally, Franklin made temperature tables for sailors to use. He urged captains of all kinds of boats and ships to take water and air temperatures to find out if they were in or out of the Gulf Stream. He asked that they share their knowledge of the boundaries of the Gulf Stream with others.

Because of Benjamin Franklin, the thermometer became a useful and popular tool of navigation. It helped ships to sail in the warm Gulf Stream when traveling northeast and in the cold ocean waters when traveling southwest. Franklin's work led to the first practical exploration of the Atlantic Ocean.

# Superstition to Satellite

As people learned about the Gulf Stream, they wondered where it came from. Using a mixture of knowledge and superstition, they tried to find the answer.

Searching for the Source

Perhaps, they thought, the Mississippi River and all other rivers that empty into the Gulf of Mexico joined together to become the big river in the ocean. It would take many rivers to create this large volume of water.

Perhaps a huge mountain of water formed each day at the equator and flowed toward the North Pole. Along the way, it created the Gulf Stream and other currents in the North Atlantic Ocean. The buildup of such a watery mountain would explain the warmth of the water and the direction it flowed. But where did the water come from each day to form this huge mountain?

Maybe the Gulf Stream welled up out of big

cracks in the bottom of the Gulf of Mexico. Such reasoning made sense to some people because they realized that the water in the ocean must **circulate**. Exactly how and where the circulation took place remained unanswered.

Some believed it took place between the surface of the earth and the depths of the earth. They were convinced that water was swallowed up by a large hole inside the earth and that it came out again as springs and rivers that flowed back to the ocean. Some believed that this great hole, or drain, pulled ships underground where they would remain forever. Any ship in the neighborhood of the drain was doomed, they reasoned, because it could not avoid such a strong pull from within the earth.

Pirates and Treasure

While some people thought about the source of the Gulf Stream, more and more ships' captains, armed with thermometers, learned to locate the big current and use it to their advantage. One group of these captains—the pirates—was known and feared by the other ships on the high seas.

Centuries ago, explorers from Europe sailed west

Winslow Homer's painting, The Gulf Stream, *reflects the fears of many sailors. In it a sailor is marooned in a ship damaged by a storm and surrounded by sharks.*

to the New World seeking gold and other valuables. Treasure ships sailed back to Europe loaded with riches.

Ships sailing in the Gulf Stream had always encountered problems from wind, weather, and the swift current. Now, ships loaded with valuable merchandise often met with another problem—pirates.

These ocean bandits sought not only gold and silver, but any valuables they could sell. A shipload of codfish was as precious as other riches. A lighter pirate ship, aided by the Gulf Stream's current, could overtake a heavily laden merchant ship sailing outside of the current. Pirates boarded these ships and plundered their rich cargoes. Piracy in the Gulf Stream became such a threat that no ship was safe from it.

Near the middle of the nineteenth century, **international law** brought an end to piracy. Increased safety for ships caused some people to take an interest in the Gulf Stream.

Matthew Maury to the Rescue

Matthew Maury, a navy lieutenant, was particularly interested. He asked sailing captains to keep daily records, or logs, of currents, air and water temperatures, and direction and **velocity** of winds. Though the sailing captains were not eager to take part in this project, they agreed to keep the records. Matthew Maury assembled and published these logs and made wind and current charts for the benefit of everyone.

His charts soon proved their value when a storm

disabled the steamship _San Francisco_. It drifted help-
lessly in the North Atlantic Ocean. No one knew
where to search for it. The last report several days
before showed the ship in the Gulf Stream 300 miles
(500 kilometers) off the coast of New Jersey. Matthew
Maury studied his charts, **calculated** the speed and
direction of the Gulf Stream's current, and marked on
the map the place he believed the ship would be found.
The rescue crew sailed there and brought back survi-
vors. The ship had sunk at the exact spot in the Gulf
Stream Matthew Maury had indicated.

The people who had doubted that his charts were
of any value now became eager to use them. Using his
charts, sailors set many new speed records. Matthew
Maury, who became known as the "father of **oceanography**," continued his studies of the ocean and published the first modern textbook on oceanography.

Now the Gulf Stream began to be **systematically**
and scientifically investigated. The United States
Coast Guard began a survey using ships specially
equipped for this job. Coast guard crews took many
measurements of the swift current at the Grand
Banks and learned more about the Gulf Stream's volume and temperature. Researchers made charts of bot-

Specially equipped coast guard ships carried out a survey to learn more about the Gulf Stream.

tom **topography** (surface features), currents, and tides.

Lieutenant J.E. Pillsbury took deep-depth current measurements. He measured for the first time the cold current that forms the riverbed of the Gulf Stream.

Oceanography and the Gulf Stream

Interest in the Gulf Stream and in oceanography has continued in the twentieth century. The Woods Hole Oceanographic Institution in Massachusetts specializes in the study of the Gulf Stream. Its **oceanographers** have measured the Gulf Stream's temperature in thousands of places. They are studying eddies—some more than a hundred miles in diameter—that break off from the main flow of the stream. They are learning more about the swift current's volume, direction, and source. Woods Hole oceanographers are also studying internal waves that break deep within the Gulf Stream rather than at its surface. These underwater waves may have been the basis of old legends about the ocean "grabbing" and "holding" ships. Through the work of the Woods Hole scientists, much valuable information is being added to our knowledge of the great ocean river.

In this century, too, oceanographers have learned

Many fish such as the school of tuna shown above live in the Gulf Stream. Since fish move when the Gulf Stream moves, it is important to the fishing industry to know exactly where the stream flows each day.

more about fish and other sea creatures of the Gulf Stream. Fishing is affected by its movement because fish stay in temperatures to their liking. When the Gulf Stream moves, they move, too.

Because the Gulf Stream affects weather, **meteorologists** are also studying it. They are learning more about how it distributes its warmth in the North Atlantic Ocean.

The View from Space

In 1976, a **satellite**, *NOAA*-5, was launched. It views the entire earth twice daily and transmits information from outer space to tracking stations on earth. *NOAA*-5's **heat-sensitive** instruments measure water temperatures in the Gulf Stream and other ocean waters. Computers make pictures of the information. Each ocean temperature shows as a different color. Using these pictures, scientists can pinpoint the exact location of the Gulf Stream each day.

Today, ships depend on satellite information to chart their daily courses. Using such accurate information, they can travel in the swiftest part of the current. This area is usually twelve to twenty-five miles (twenty to forty-two kilometers) from the western edge of the stream. Following this course saves a considerable amount of fuel because the warm current helps move ships through the ocean.

The Gulf Stream is an important and complex current. Our knowledge about it keeps changing. Some of our present "facts" may change as we learn more about this mighty stream. Many oceanographers are finding challenging careers in their work on this new and exciting scientific frontier.

A Submarine Investigates

Two hundred years after Benjamin Franklin named the Gulf Stream, a research submarine—named *Ben Franklin* after the famous American—made a historic journey in the great ocean river. The *Ben Franklin* had been built particularly for this voyage and was equipped with all the scientific instruments necessary to accomplish its mission.

The Mission of the Ben Franklin

The mission of the big yellow and white submarine was to drift in the Gulf Stream at certain depths for thirty days. Using no power of its own, it would move along at the same speed as the current. During the drift, the submarine's crew would plot the course of the great ocean river along the coast of North America, learn its speed, and study plant and animal life in the stream. It would be the first mission of its kind.

Dr. Jacques Piccard, a Swiss oceanographer, headed the underwater expedition. Other crew members of

The Ben Franklin *as it was launched for its historic journey in the waters of the Gulf Stream.*

this historic mission were a captain, a submarine pilot, two more oceanographers, and a **NASA** engineer.

The Research Begins

On July 14, 1969, near Palm Beach, Florida, the *Ben Franklin* settled beneath the water's surface near the center of the Gulf Stream. Because this spot is fairly shallow, the submarine hovered only 32.8 feet (10 meters) from the ocean floor.

As soon as the submarine began drifting in the powerful current of the ocean river, its **research** began. Every two seconds a Water Analysis Sensing Pod registered the water's temperature, its **salinity** (saltiness), the speed of sound in this part of the Gulf Stream, and the depth of the submarine. At the end of its journey, nearly four million measurements would be fed into a computer. All that information would be studied by scientists to gain valuable knowledge about the Gulf Stream.

The *Ben Franklin* was large enough for members of the expedition to be fairly comfortable even though it also contained much scientific equipment. Unlike most submarines, which have no portholes, the *Ben Franklin* had twenty-eight windows on the sea.

Crew members took turns sleeping so that at least two were awake and on duty at all times. Their research continued day and night so that nothing of importance to the mission would be missed.

While the others studied the Gulf Stream, the NASA engineer watched the other men closely. He wanted to learn more about the needs and behavior of people who are isolated for weeks at a time. This information would be useful in the future for scientists living in space stations.

A Lifeline to the Outside World

On the swift current's surface, *Privateer*, the escort ship, stayed in the same area as the *Ben Franklin*. Throughout the journey, the two vessels kept in contact with each other. They sent messages back and forth by an underwater telephone.

Every half hour the *Privateer* called from the surface, "*Ben Franklin*, this is *Privateer*. Over."

The *Ben Franklin* answered, "Roger. Out." In order to save electric power, unless a further message was necessary, nothing more was said for another half hour. In its **isolated environment** deep in the Gulf Stream, the *Ben Franklin* depended on the *Privateer*

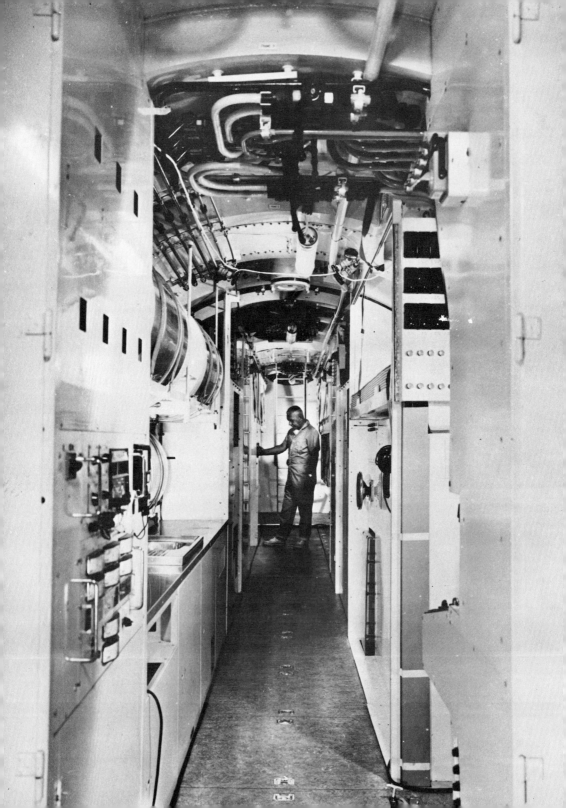

for everything its crew members needed to know.

It didn't take long for the mission to encounter its first problem. The Gulf Stream flows faster at the surface than below the surface where the *Ben Franklin* drifted. The *Privateer* kept getting ahead of the *Ben Franklin*. In order for the escort ship to remain even with the submarine, it had to sail slowly backwards for the entire 1,500-mile (2,500-kilometer) trip!

Windows on the Sea

Through the portholes, the submarine's crew watched and photographed sea creatures of the Gulf Stream. Tiny **copepods**—the food of many sea animals—passed in front of the portholes. **Transparent**, long moving chains of **salpas** moved by, too. Sometimes the salpas rolled into spirals and drifted in the water.

Soon the oceanographers tested another new idea. A large chamber, or **plankton** tube, had been built into the submarine to trap tiny sea creatures. Here they moved about, unaware that they were captives or that they were being studied. Then a valve was opened, freeing the creatures unharmed. The plankton tube was a success.

This is a view of the inside of the Ben Franklin. *During its mission two crew members were on duty at all times.*

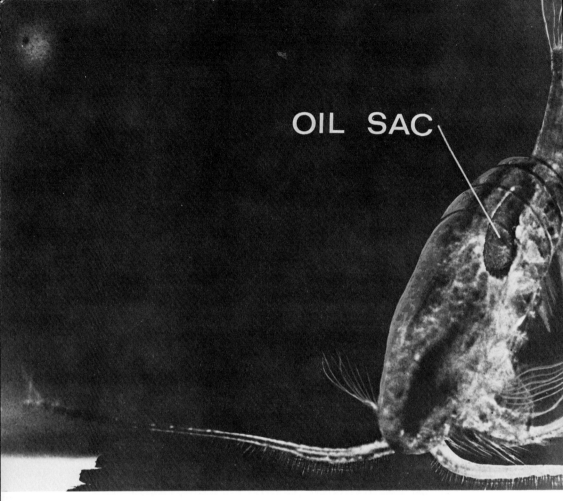

OIL SAC

Tiny copepods—the food of many sea animals—passed in front of the portholes of the research submarine. This highly magnified photograph of a copepod was taken through a powerful microscope.

Crabs and eels passed by the portholes, as did shrimp, arrowworms, and jellyfish. The crew watched **diatoms**, tiny creatures that sparkle in the water. They saw rat-tail fish, sea urchins, tuna, and barracuda. When several large sharks passed the portholes and a hammerhead shark circled the submarine a few

times, the *Ben Franklin* used the underwater telephone to warn the *Privateer*. If divers from the ship
were out, they would need to know that sharks were
nearby.

Since the *Ben Franklin* drifted quietly along, most
fish paid no attention to it. Some gathered at the portholes when floodlights were turned on. One fish, however, seemed to dislike the submarine and considered
it to be an enemy.

Early one morning while three of the crew were
awake, a beautiful five-foot-long broadbill swordfish
swam back and forth close to the portholes. It seemed
to be trying to get a better look at the submarine. The
swordfish swam away; then it returned. Suddenly it
attacked, striking the submarine's hull with the point
of its sword. The impact was clearly heard by three
crew members. The submarine was not harmed. Ap-
parently the swordfish decided this enemy was too
big and strong, and it swam away.

Sometimes the *Ben Franklin* drifted at 600 feet
(184 meters) below the surface of the Gulf Stream.
Several times, in order to do research nearer the ocean
floor, it descended to about 2,000 feet (614 meters)
below the surface. In these lower depths the temperature

At one time a hammerhead shark circled the Ben Franklin. *The crew used its underwater telephone to warn the* Privateer. *(Robert Commer/ EarthViews)*

inside the submarine cooled below 53.6° F (12° C). To keep warm, the men exercised. After they took photos of Gulf Stream creatures through the portholes, they warmed their hands on the flashbulbs.

Eddies and Hurricanes

For eleven days all went well with the *Ben Franklin*. On the twelfth day, the submarine was caught in a huge eddy. The crew could only watch as their submarine was pushed out of the Gulf Stream and moved seventeen miles (about twenty-eight kilometers) west of the stream.

Since the *Ben Franklin* could not get back into the Gulf Stream by itself, it rose to the surface and was towed back by the *Privateer*. Then it again descended into the stream and continued its mission.

Several days later, the *Privateer* alerted the *Ben Franklin* that the season's first **hurricane**—Hurricane Anna—was forming. Its position was about 93 miles (150 kilometers) southwest of the surface ship.

Would Hurricane Anna follow the Gulf Stream as many hurricanes do? If so, the *Privateer* would be in great danger.

The crew had agreed earlier that, in case of an

When the Ben Franklin *was pushed out of the Gulf Stream by an eddy, it had to surface to be towed back on course by the* Privateer.

emergency, the *Ben Franklin* would surface and be towed by the *Privateer*. The *Privateer*'s speed would be slow while towing the submarine. If the hurricane followed the Gulf Stream, it might overtake them.

The storm's course could not be accurately predicted. Perhaps it would die out. Perhaps it would change direction. If the *Privateer* and the *Ben Franklin* left the Gulf Stream now, they could not learn what they had set out to learn about the swift current.

The two ships decided to wait and watch Hurricane Anna a little longer. It was the right thing to do. The storm soon vanished from the area.

Mission Accomplished

On the thirtieth day of its journey in the Gulf Stream, the *Ben Franklin* surfaced near Halifax, Nova Scotia. Its pioneering mission was over.

During the 1,500-mile (2,500-kilometer) drift in the Gulf Stream, the submarine's crew had developed a new method of observing the ocean at different depths. They had plotted the course of the Gulf Stream on the swiftest part of its journey and had charted its speed at various depths. They had successfully tested the plankton tube and had acquired new knowledge about sea creatures. Oceanographers aboard had measured **chlorophyll** and mineral content of the swift current. The submarine's **gravimeter** had measured gravity in the Gulf Stream, and a **photomultiplier** had gathered information about the transmission of light. Hundreds of photographs had been made of the ocean floor. All this information added to our knowledge of the Gulf Stream.

The mission of the *Ben Franklin* had answered many questions about the Gulf Stream and raised many more questions. It had provided valuable information about how people behave in an isolated, alien environment. And it had established a new approach to studies in oceanography.

6 When Heat Meets Cold

The **Grand Banks**—the area near Newfoundland where the warm Gulf Stream clashes with the cold Labrador Current—covers thousands of square miles. The results of this tremendous collision are sometimes harmful to people and sometimes helpful.

Because the Gulf Stream water is warm, the air above it is warm. Likewise, because the Labrador Current is cold, the air above it is cold. When warm air and cold air meet, fog can result. One of the densest fogs anywhere in the world is produced at this spot above the Gulf Stream. **Visibility** is so poor that ships have run together and lives have been lost.

Beware of Icebergs

Icebergs add to the problem at the Grand Banks. Each spring the sun shines on Arctic glaciers, causing huge chunks to break off. These chunks, or icebergs, drift south in the Labrador Current. Often they enter the Gulf Stream. Some of the thousands of small ice-

U.S. Coast Guard ships help patrol the shipping lanes and keep track of dangerous icebergs.

bergs floating in the warm current melt almost at once. Large ones take a week or more to melt.

Icebergs present a special danger because they lie mostly beneath the water's surface where their undersea part cannot be seen. In the fog at the Grand Banks, ships and icebergs have collided.

As an iceberg drifts in the Gulf Stream, warm water eats away at its outer surfaces, sharpening its edges. To understand how these sharp edges appear, hold one side of an ice cube under warm, running water for a minute or so. Watch how the warm water melts parts of the ice cube, leaving irregular, jagged edges.

In 1912 the sharpened edge of an iceberg floating in the Gulf Stream caused a tragedy. It tore a hundred-yard gash in an ocean liner, the *Titanic*. This huge ship was the largest in the world at that time. Because it was thought to be unsinkable, not even the ship's crew fully understood the danger of their situation. Suddenly, less than three hours later, the *Titanic* plunged into the ocean depths, killing more than 1,500 passengers and crew.

Since this terrible accident, the danger from icebergs in the Gulf Stream and elsewhere has been greatly reduced. The International Ice Patrol now warns ships of the presence of icebergs and alerts them to the icebergs' locations. When icebergs get into **shipping lanes**, often they are broken up and destroyed by the ice patrol.

A "Silver Mine" for New England

The clash of warm Gulf Stream water and cold water of the Labrador Current also causes undersea turbulence. Beneath the surface, the water is constantly churning from top to bottom. As the ocean floor is stirred up, **nutrients** are brought to the surface. This constant supply of nutrients in the Gulf Stream creates a **food-chain reaction** that benefits the northeastern part of the United States.

Plankton, the ocean's **microscopic** plant and animal life, thrives on these nutrients. Plankton floats, depending on movement of the water for speed and direction. With a constant supply of nutrients, plankton in this part of the Gulf Stream has grown into large floating pastures of greenery.

Because plankton is abundant, fish are attracted to this Gulf Stream area in larger numbers than anywhere else in the ocean.

Plankton feed on the nutrients that are brought to the surface where the Gulf Stream runs into the Labrador Current. In this photo, zooplankton—tiny animals in the sea—are shown enlarged many times their normal size.

"Banks" are names given to parts of the ocean where there are many fish. The "Grand Banks" is well named, for billions of fish live here. This turbulent Gulf Stream area has been called a "silver mine" because of its great numbers of silvery-colored codfish. Norse explorers discovered the Grand Banks before America was settled. Now it is one of the most popular fishing spots in the world.

Early settlers in North America discovered that the silvery codfish had special qualities. Unlike most fish, codfish did not have to be eaten immediately. They could be preserved by salting and smoking. Then they could be stored or shipped long distances without refrigeration.

A female codfish lays millions of eggs each year. Though most eggs and young codfish become food for larger fish, many others grow to adulthood. To the early settlers, the codfish supply seemed limitless.

Because of the Gulf Stream's clash with the Labrador Current, great fisheries developed in Maine and Massachusetts. Fish were taken to shore where they were cleaned, laid out in the sun to dry, and preserved by salting and smoking. When they were shipped to European markets, the Gulf Stream served as a water highway to transport them.

The codfish industry provided jobs for many people, and the economy of these states thrived. Because New England seemed to be a good place for newcomers to settle, many came. Boston became the "Codfish Capital" of the world. The Massachusetts House of Representatives passed a resolution recognizing the importance of the codfish. The lawmakers

Green sea turtles and many other sea creatures are attracted to the Grand Banks by the abundant food supply. (Ken Howard/EarthViews)

hung an emblem—a large gilded pine codfish—in the room where they met. Today the emblem is still there.

Giants of the Sea

In addition to cod, other fish and sea creatures are lured to this part of the Gulf Stream by the abundant food supply. Sea turtles and ocean birds come to find tasty meals. Great white whales are attracted because enough plankton is here to satisfy their huge appetites. With mouths open, they cruise through a plankton meadow at the edge of the Gulf Stream. By straining the plankton through their **whalebone jaws**, they retain the food but not the water.

Long ago, whalers knew the whale's feeding habits and its preference for cold water. They also knew that large pastures of plankton in the Gulf Stream attracted whales. As a result, for centuries whale hunting flourished at the Grand Banks. Finally, most of the whales had been killed. Today, a group of nations has made whaling in the Gulf Stream illegal. Once more these giants of the sea come to feed along the edges of the great ocean river.

7 Palm Trees and Eels

As the Gulf Stream transports heat from the tropics, winds that blow across this warm current absorb some of its heat and carry it to nearby land. In the Northern Hemisphere, westerly winds are the prevailing, or dominant, winds. In addition to helping move the Gulf Stream, they carry Gulf Stream heat to countries north and northeast of the warm current. These countries benefit greatly from this warmth. Because the sun's heat absorbed by the Gulf Stream penetrates far below the water's surface, it can be transported long distances by the current.

As moisture-filled westerlies blow over nearby land, temperatures of the land become more moderate. Large bodies of water make cold land warmer and hot land cooler. This happens because water releases heat absorbed from the sun slower than land releases it. Unlike inland desert areas that release heat rapidly and have extremes of hot and cold, temperatures of land near oceans are more nearly equal.

A Warm Current in the Far North

Iceland is one country that benefits from the Gulf Stream warmth. Iceland borders the extreme cold of the Arctic Circle. Snowfields and more than a hundred glaciers cover parts of this country. Even so, the climate is moderate because the Gulf Stream flows near the country's southern coast. Winds blowing toward Iceland absorb heat from the powerful stream and carry it inland.

Because of the warmth from this current, cattle and sheep graze in Iceland's high pastures each summer. Grass and corn grow in the fertile valleys. Winter temperatures here are warmer than winter temperatures of Maine or Vermont. Since these New England states are much farther south, they might be expected to be warmer than Iceland.

The Gulf Stream also transports heat to Norway, a spoon-shaped country that extends into the Arctic Circle. Eastern Norway benefits very little from the magic of the Gulf Stream because winds do not carry heat that far inland. Western and southern Norway, however, have cool summers and mild winters.

Hammerfest, a Norwegian fishing town and the northernmost town of Europe, lies well within the

Because of the warm waters of the Gulf Stream, pasture lands and fertile farmlands cover many areas in Iceland. This small, volcanic country borders the Arctic Circle in the North Atlantic.

Other parts of Iceland have snow-covered mountains and rugged areas covered by icy glaciers. Even here the climate is warmer than it would be without the influence of the Gulf Stream.

Arctic Circle. Because of warmth from the Gulf Stream, the port of Hammerfest is open during all seasons of the year. Seaports hundreds of miles south in regions not warmed by the Gulf Stream are ice-bound in winter.

The British Isles, too, are located in the far north—about as far north of the equator as Siberia. Yet, because of the warm winds from the Gulf Stream, Britain's temperatures are much more moderate. It has been said that the Gulf Stream is Britain's best friend. It is responsible for the lovely trees and a wealth of flowers that grow in the British Isles. Roses bloom until December, and grass is green all year. Hedge fences are common. Wildflowers and ferns cover poorer soil. Because of the Gulf Stream's warmth, cattle feed in the green pastures. Great Britain even has a few palm trees—warm weather, semi-tropical plants.

The tiny islands of Bermuda, located straight east of North Carolina, are truly blessed by the Gulf Stream. While inland regions of the United States bake in summer heat, Bermuda has warm, comfortable temperatures. While the same inland areas battle cold weather in winter, Bermuda never has frost. No

The mild climate and lovely trees and flowers of the British Isles exist because of the nearby warm waters of the Gulf Stream.

part of Bermuda is farther than a mile from the ocean. Warm air from the Gulf Stream blows over all of Bermuda's beaches, bamboo, and banana trees. No wonder these islands appear to be one big flower garden and are ideal vacation spots!

A Highway for Sea Creatures

In addition to transporting heat, the Gulf Stream is a highway for many sea creatures. One of these, the eel, is a food highly valued in parts of Europe.

Eels lay their eggs in the thick weeds of the Sargasso Sea. As the eggs hatch, the tiny creatures drift into the Gulf Stream. For the next three years, they drift with this mighty current for thousands of miles. Then, all of a sudden, they head for land.

On European shores at springtime, the tiny creatures battle currents and tides and finally wiggle into freshwater streams. Transparent except for their eyes, these finger-length eels are called elvers. Much of their growth comes after they leave the Gulf Stream and enter freshwater streams.

The Europeans know approximately when eels will arrive and catch many of them. Eels that escape will, twenty years or so later, again enter the Gulf

Stream. Drifting in its current, they will migrate back to the Sargasso Sea to spawn in their original saltwater home.

Although the Gulf Stream flows near land all the way up the east coast of the United States, most of the coast receives very little warmth from it. One reason is that westerly winds blow warm air the opposite direction—into the ocean. Another reason is that a wall of cold water flows southward between the Gulf Stream and the coastline.

While the United States receives little warmth from the Gulf Stream, it does gain other benefits. We have seen how the codfish industry became important in the New England states because of the Gulf Stream. Since many kinds of fish that prefer warm water live in the stream, fishing off the east coast of the United States flourishes. Dolphins live there, too. These intelligent mammals are valuable helpers in underwater exploration of the Gulf Stream. Some have been taught to carry supplies to underwater crews, to locate lost tools, and to aid divers.

The Gulf Stream brings one creature to coastal areas that is not so welcome—the Portuguese man-of-war. Its long, stinging tentacles hang down into the

water. Above the surface, a pinkish sail allows the wind to move the Portuguese man-of-war along. When storms cause large waves to pound the shores, beaches sometimes must be closed to swimming be-cause of these creatures. Unlike the eel, no one wants to eat them.

Dolphins are intelligent, playful, and graceful mammals that live in the Gulf Stream. (Robert Pitman/EarthViews)

8 The Weathermaker

Oceans have a great influence on weather over both sea and land areas. Because the warm Gulf Stream is the dominant current of the North Atlantic Ocean, it influences weather in the ocean and in many North Atlantic land regions.

The sun's rays heat land and water. Since these rays penetrate only the land's surface, this daytime heat is quickly lost at night. In oceans, though, the sun's rays penetrate hundreds of feet below the water's surface. This heat is stored and then released slowly, day by day, summer and winter.

Stormy Weather

If air temperatures and moisture levels were the same everywhere, there would be no storms. Changes in weather occur when air currents of different temperatures and moisture levels meet. Over the oceans, warmer air clashes with colder air. Since the Gulf Stream's temperature is higher than that of other ocean water, clashes above the Gulf Stream are more severe than clashes above other parts of the ocean.

Sometimes fog is produced when warm and cool air clash. We have seen that warm air above the Gulf Stream clashes with cold air above the Labrador Current to create dense fog over an area of 150,000 square miles. These fog banks sometimes reach more than 1,000 feet (305 meters) into the air.

Fog is also created as warm Gulf Stream air collides with cooler air on land. In England, London is noted for its fog. At times the east coast of the United States has fog which occurs when local winds bring the warm air inland.

In addition to fog, this clash of warm and cool air produces rain, snow, and powerful winds. All parts of the Gulf Stream are very stormy and rough because collisions frequently take place between warm and cool air. Early sailors dreaded storms here more than in any other part of the ocean. They spoke of the "ugly" sea raised by such sudden storms.

A Highway for Hurricanes

A hurricane is one of the most violent kinds of weather related to the Gulf Stream. It is spawned in this warm current as moist air rises and begins whirling in a counterclockwise direction. Under certain conditions, this whirling air mass becomes a **tropical storm**. When the winds reach a speed of 74 miles per hour (119 kilometers per hour), the storm becomes a hurricane. Then it is driven by its own energy.

A hurricane is one of nature's deadliest bundles of fury. It is a triple threat, causing damage from strong winds, high waves, and cloudbursts of rain. It can develop in any ocean area where there is extremely warm, moist air.

The Gulf Stream, because of its warm water and the warm air above it, is a kind of highway for hurricanes each summer. Ships move out of the area when hurricanes form. People living in nearby coastal towns leave their homes or take steps to protect themselves and their property. Although warnings are issued well in advance of the approach of a hurricane, weather forecasters cannot predict its exact course, speed, and strength. A hurricane may lose some of its strength, and then regain it from the Gulf Stream's warm water. It may travel for eight to twelve days. Finally, when its energy is spent over the ocean or on land, it loses its awesome destructive power.

Photographed from a satellite in space, Hurricane Diana appears as a huge, whirling mass of clouds. In this picture it is centered just off the coast of North Carolina.

Indirectly, the Gulf Stream also spawns **tornadoes**. As warm air moves inland with a hurricane, it may clash with a cold front. A tornado develops when the warm and cold air wrap around each other and form a violent, twisting funnel. When a tornado strikes land, it often destroys everything in its path.

⑨ Solving Today's Problems

Can the Gulf Stream, along with oceans everywhere, help feed the world's exploding population? Many scientists believe it can.

Researchers are searching for ways to obtain more and better food from the Gulf Steam and all oceans. They are looking for a food that can be stored and can be shipped long distances without refrigeration.

Food from Seaweed and Plankton

Some scientists are looking at plant life. Seaweed grows in many places in the oceans, including the Gulf Stream. Japan and several other countries use it as food. Some seaweed is rich in vitamins and iodine, substances that humans need to stay healthy. Scientists are trying to find the best seasons for harvesting seaweed and the best ways of gathering it. They are searching for ways to prepare seaweed so that it will be tasty.

Seaweed, such as this floating mass near Cape Cod, Massachusetts, may provide a rich source of food in the future. (Lynn Stone)

Another form of plant life in the Gulf Stream is phytoplankton. It drifts in the water and is the "grass" of the sea upon which many sea animals feed. Because it is microscopic in size, phytoplankton is difficult to collect. Some of it is so tiny that it can pass through the finest gathering net. While the Gulf Stream has a good supply of phytoplankton, better, inexpensive ways to harvest it need to be developed.

Hungry people need **protein** in their foods. Fish and other sea creatures are rich in protein.

Zooplankton, the floating animal part of plankton, is made up of tiny animals that feed on the phytoplankton. Some are shrimplike. Many are no longer than the head of a pin. Each has a shell. Before they can be used as food, though, a good method of removing the shells from these tiny creatures must be found.

Some larger sea animals in the Gulf Stream could be used as food. Their flesh is rich in protein, but many people would not like to eat them. For example: Would you eat squid for dinner? It is eaten and enjoyed by people in some countries.

Sometimes people reject food because they have never tasted it or because its name or appearance is not pleasing. Before these foods can help solve the

The squid shown here are one example of many sea creatures that could help solve the world hunger problem. Squid are rich in protein and other nutrients.

world hunger problem, new ways must be found to prepare them. If they look attractive, then people will try them and discover how tasty they are.

When decisions are made about fishing and whaling in the Gulf Stream and other ocean areas, **conservation** should play an important role. Removing too many of any species could cause it to become **extinct**. Also, **depleting** the food supply of any sea animal may put it in danger of extinction. Ocean life can be preserved and protected for future use at the same time it is being used to supply food for people today.

Fish for a Hungry World

Fishing in the Gulf Stream has been important for centuries. In order to increase fishing, new fishing methods need to be developed. For example, electrical devices track schools of fish. Electrical fields can keep fish penned until they are needed. Other methods are being tried or considered.

Many people like the taste of such Gulf Stream fish as cod and herring. Large numbers of these fish have already been used as food. To conserve them, we must know how many can be caught each year without placing them in danger.

Scientists have another idea to increase the amount of food from the sea. They want to make use of fish now thought of as "trash" fish. In the Gulf Stream there are many trash fish. Some are thrown away when they are caught in nets along with fish people like to eat. Others are now being made into a powder that is used as food for animals. Since fish powder is rich in protein, animals thrive on it.

Many "trash fish" now caught in the Gulf Stream could be used to make fish flour and help feed hungry people.

Flour is also made from trash fish. When a fish flour is added to wheat flour, the combined product can be used in baking bread, cookies, and cakes. This enriched flour adds valuable protein to baked goods and can be used almost anywhere wheat flour is used.

Fish flour, known as FPC, or fish protein concentrate, is now sold in some countries. FPC can be stored and can be shipped long distances without refrigeration. Some scientists believe FPC is the best approach to the world hunger problem. They believe that we will find ways to produce FPC cheaply. Then it can be shipped wherever it is needed to feed the world's hungry people.

Energy from the Gulf Stream

Another world problem is the shortage of **energy** and the **natural resources** needed to produce it. Coal, oil, and natural gas are used in power plants to produce electric energy. These resources are expensive and are being depleted rapidly. Scientists are searching for ways to produce more and cheaper electricity.

Some scientists believe the Gulf Stream can help solve this energy problem. Imagine **turbines** placed 1,000 feet (305 meters) down in the Gulf Stream off the coast of the United States. The steady, rapid flow of the Gulf Stream would turn the turbines every day and every night, year after year. The turbines would turn above-water **generators** that would produce electricity. Such a system would not **pollute** the water or cause great changes in the Gulf Stream's temperature. The electricity produced could be furnished to states on the east coast of the United States.

Many energy sources can be used only once. When an oil well is depleted, oil must be found somewhere else. The same is true with coal. The Gulf Stream, however, is a **renewable** energy source. It will never be used up no matter how many turbines are placed in it. Water used to turn the turbines will remain in the Gulf Stream. Electrical power can be produced year after year without depleting the energy source. Because of this, electricity produced should be inexpensive.

Another world problem is overcrowding of people in certain areas. The Gulf Stream may help solve this problem, too. Some scientists believe that floating cities or underwater towns may be practical ideas. If these ideas are actually used, the Gulf Stream may become an underwater highway for people.

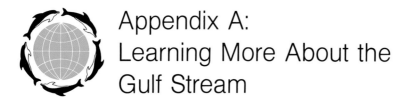

Appendix A:
Learning More About the
Gulf Stream

The following activities will help you to learn more about the Gulf Stream. Choose one or more to begin working on today.

1. Draw a map of the Atlantic Ocean. Mark with arrows where the Gulf Stream flows. Start at the Gulf of Mexico and show how the Gulf Stream squeezes through the Straits of Florida and moves up the Atlantic coast. Locate Cape Hatteras, North Carolina. Show with double arrows how the ocean river begins to widen and moves northward a little farther from shore as far as New England. At Newfoundland, let the arrows move in circles to show the turbulent waters, and then move on clockwise across the Atlantic Ocean, showing that the Gulf Sream becomes wider as it goes. Let the arrows show the stream dividing into branches as it flows southward past Europe. Locate the Sargasso Sea. Show the Gulf Stream circling clockwise around it before moving on south toward the equator.

2. Read a biography of Benjamin Franklin. Does it tell about his using a thermometer to test water temperature in the Gulf Stream? What else did he do that added to our knowledge of the Gulf Stream?

3. Start a notebook about hurricanes and tornadoes. How many of the hurricanes were spawned in the Gulf Stream? What time of the year did most of the hurricanes occur? Could some of the tornadoes have been spawned by hurricanes that started in the Gulf Stream? Cut out newspaper accounts of these storms and glue them in your notebook. Learn what precautions you should take when a hurricane or tornado approaches.

4. Go to a meat market or the meat department of your supermarket. Count how many kinds of fish and other sea creatures are there. Write down the name of each one. Read the labels on the packages of frozen fish and seafood to learn how many of them came from areas near the Gulf Stream. Ask the butcher if some of the fresh fish and seafood came from the Atlantic Ocean.

 Appendix B:
Scientific Names
for Sea Animals

Sea creatures, like all living things, have two kinds of names. The first is their *common name*, a name in the everyday language of an area where they are found. An animal often has a number of different common names in different languages. Also, several different animals may be known by the same common name.

The second kind of name is their *scientific name*. This is a Latin name assigned by scientists to identify an animal all over the world for other scientists. The scientific name is usually made up of two words. The first identifies a genus, or group, of similar animals (or plants), and the second identifies the species, or kind, of animal in the group. Sometimes, as scientists learn more about an animal, they may decide it belongs to a different group. The scientific name is then changed so that all scientists can recognize it and know exactly what animal it refers to.

If you want to learn more about the creatures in this book, the list of scientific names that follows will be useful to you. A typical species has been identified for each type of animal mentioned in the book. There may be many other species in the same group.

Chapter	Common Name	Scientific Name
2.	Flying Fish	*Exocoetus volitans*
	Atlantic Sea Horse	*Hippocampus hudsonius*
	Blue Crab	*Callinectes sapidus*
4.	Codfish	*Gadus callarias*
5.	Copepod	*Cyclops calanus*
	European Eel	*Anguilla anguilla*
	Shrimp	*Crago vulgaris*
	Arrowworm	*Sagitta bipunctata*
	Sea Urchin	*Echinus esculentus*
	Tuna	*Thunnus thynnus*
	Barracuda	*Sphyraena barracuda*
	Shark (Atlantic Mako)	*Isurus oxylhinchus*
	Hammerhead Shark	*Sphyrna zygaena*
	Swordfish	*Xiphias gladius*

Chapter	Common Name	Scientific Name
6.	Green Sea Turtle	*Chelonia mydas*
	White Whale	*Delphinapterus leucas*
7.	Common Dolphin	*Delphinus delphis*
	Portuguese Man-of-War	*Physalia physalis*
9.	Common Squid	*Loligo pealei*
	Herring	*Clupea harengus*

 Glossary

Arctic Circle—a small circle of the earth that is parallel with the equator; the coldest regions of the Northern Hemisphere lie between the Arctic Circle and the North Pole

calculate—to figure by mathematical methods

chlorophyll (KLAWR-uh-fihl)—green coloring matter of leaves and plants

circulate—to move in a circle or follow a course that returns to the starting point

conservation—preserving from loss or waste

contract—to draw together

copepod (KO-peh-pod)—a tiny crustacean having a hard shell or crust

Coriolis (kawr-ee-OH-lihs) effect—the earth's spin, which causes the Gulf Stream and other moving objects to be curved rather than straight

density—the amount of a substance contained in a given volume, area, or length

deplete—to decrease or exhaust the supply

diatom (DY-uh-tahm)—a tiny, one-cell algae with hard, glasslike cell walls that seem to sparkle in seawater

eddy—a whirling current; giant, powerful eddies whirl away from the main current of the Gulf Stream and often surround a pocket of cold water

energy—the ability to be active or do work; many forms of energy are used to meet human needs

equator—an imaginary line circling the earth at equal distances from the north and south poles

expand—to increase in volume

extinct (ek-STINKT)—no longer living anywhere on earth

food-chain reaction—small animal or plant life becomes food for larger creatures which become food for still larger creatures, creating a support system that links the smallest form to the largest

generator—a machine that changes one form of energy into another; generators are often used to produce electricity

Grand Banks—a foggy ocean area southeast of Newfoundland where the warm Gulf Stream clashes with the cold Labrador Current

gravimeter (grah-VIM-it-er)—an instrument for measuring the force of gravity

Gulf Stream—a warm ocean current flowing north from the Gulf of Mexico that merges with the North Atlantic Current near Newfoundland

heat-sensitive—used here to describe instruments that react to heat

hemisphere—half of the globe; the earth has a northern and a southern hemisphere

hurricane—a tropical, cyclonic storm accompanied by rain, thunder, and lightning with winds of at least seventy-four miles per hour

iceberg—a large, floating mass of ice, detached from a glacier and carried out to sea

international law—a set of rules that control or affect the rights of nations in their relations with each other

isolated environment—a place or living situation separated from other persons and familiar things

Labrador Current—a cold current flowing south from the Arctic Circle that collides with the Gulf Stream at the Grand Banks and then flows beneath it

meteorologist (mee-tee-uh-RAHL-uh-jist)—a scientist who studies the earth's atmosphere, especially its weather and climate

microscopic (my-kro-SCOP-ik)—extremely small in size; an object that can be seen only with a microscope

NASA—National Aeronautics and Space Administration, the government agency in charge of the U.S. space program

natural resource—a source of supply of natural materials used to meet human needs

nutrients (NYU-tree-ents)—substances that promote the growth of living things

oceanographer (oh-shuh-NOG-reh-fer)—a person who uses scientific methods to explore and study the ocean

oceanography (oh-shuh-NOG-reh-fee)—the exploration and scientific study of the ocean

photomultiplier (foh-tuh-MUL-tuh-ply-er)an extremely sensitive detector of light and of other radiant energy in the form of waves or particles

phytoplankton (fy-tuh-PLANK-tuhn)—the tiny plants that drift in the ocean and are the "grass" upon which many sea animals feed

plankton—the mostly microscopic plant and animal life in the ocean and other bodies of water; plankton either floats in the water or swims weakly

pollute—to make foul or unclean

pontoon—a floating structure; seaweed berries in the Sargasso Sea serve as pontoons to keep the plants afloat

protein (PRO-teen)—any of the many naturally occurring combinations of amino acids that form essential parts of all living cells

pulsate—to expand and contract in rhythm

radiating heat—sending out heat

renewable—capable of being used again, as a renewable energy source

research—a systematic study

rotation—used here to mean the earth's movement or spin

salinity (suh-LIN-it-ee)—used here to mean the salt content of seawater

salpa (SAL-puh)—a tiny ocean creature with a transparent, saclike body

Sargasso (sahr-GAS-oh) Sea—a calm area of water in the North Atlantic Ocean northeast of the West Indies; plant and animal life thrives in this seaweed-covered "hot water tank"

satellite—used here to describe artificial satellites—human-made objects launched from the earth into orbit around a planet

shipping lanes—paths in which ships travel

systematic (sis-tuh-MAT-ik)—using a plan or method

thermometer—an instrument for measuring temperature

topography (tuh-PAHG-ruh-fee)—the description of the features of a natural or human-made surface

tornado—a violent, destructive whirling wind accompanied by a funnel-shaped cloud that progresses in a narrow path over land or water

transparent—made of such material that objects can be seen on the other side; see-through

tropical storm—a tropical cyclone having wind speeds of less than seventy-four miles per hour

turbine—a machine having a rotor, usually with vanes or blades, and driven by pressure of a moving fluid such as water

turbulent (TUHR-byu-lent)—moving in many directions; causing unrest or disturbing normal movement

velocity—speed

visibility—the ability to be seen under given conditions

volume—the amount of a substance occupying a particular space

whalebone jaws—an elastic, horny substance growing in place of teeth in the upper jaw

whirlpool—a strong, whirling current

zooplankton (zoh-uh-PLANK-tuhn)—tiny animals that feed on phytoplankton floating in the sea

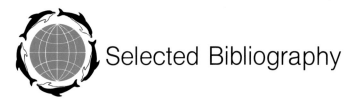 Selected Bibliography

Books

Archer, Jules. *Hunger on Planet Earth*. New York: Crowell, 1977.

Atkinson, Bruce W. *The Weather Business*. New York: Doubleday, 1969.

Bergaust, Erik, and Foss, William Q. *Oceanographers in Action*. New York: G.P. Putnam's, 1968.

Boesen, Victor. *Doing Something about the Weather*. New York: G.P. Putnam's, 1975.

Carter, Samuel III. *The Gulf Stream Story*. New York: Doubleday, 1970.

Cowen, Robert C. *Frontiers of the Sea*. New York: Doubleday, 1969.

Darby, Ray. *Conquering the Deep Sea Frontier*. New York: McKay, 1971.

Daugherty, Charles M. *Searchers of the Sea*. New York: Viking, 1961.

Hillman, Hal. *Feeding the World of the Future*. Philadelphia: Lippincott, 1972.

Idyll, Clarence P. *The Sea Against Hunger*. New York: Crowell, 1970.

Piccard, Jacques. *The Sun Beneath the Sea*. New York: Scribner's, 1971.

Soule, Gardner. *The Ocean Adventure.* New York: Appleton Century, 1966.

Articles

Ashkenazy, Irvin. "Bethnic Currents, Unstable Highways in an Arcane World." *Oceans,* November 1979.

Geary, Ida. "Seaweed Washed Ashore." *Oceans,* February 1980.

Harris, Marvin. "It's Only a Cod." *Natural History,* November 1979.

 Index

The photographs are reproduced through the courtesy of the British Tourist Authority; EarthViews (Robert Commer, Ken Howard, and Robert Pitman, photographers); the Environmental Science Services Administration; the Florida Division of Tourism; The Grumman Aerospace Corporation; the Iceland Tourist Bureau; Lynn Stone; the National Archives; the National Oceanic and Atmospheric Administration; The Scripps Institution of Oceanography; the U.S. Coast Guard; and the Wometco Miami Seaquarium. Cover: the Ben Franklin in New York Harbor.

About the Author

Alice Gilbreath is the author of eighteen previous books for young people during a twenty-year writing career. Her wide-ranging interests have led her to write about numerous subjects, from beginning crafts to the defensive techniques of animals.

"In writing this book," says the author, "I have tried to share with young people a glimpse of this unique ocean river and how it affects all of our lives. It has altered our history, navigation, and industry. The Gulf Stream has been both friend and foe, but it never is neutral. I hope that young people will be so intrigued by this marvelous current that some will consider becoming oceanographers."

Ms. Gilbreath attended Trinity University in San Antonio, the University of Tulsa, and the College of Idaho. She lives in Bartlesville, Oklahoma.